Design ideas for drones, multi- and quadcopters
Roland Büchi

Bibliografische Information der Deutschen Nationalbibliothek
Die Deutsche Nationalbibliothek verzeichnet diese Publikation in der Deutschen Nationalbibliografie; detaillierte bibliografische Daten sind im Internet über www.dnb.de abrufbar.

Impressum
Roland Büchi, 2022
Herstellung und Verlag: BoD – Books on Demand, Norderstedt
ISBN: 978-3-7557-8094-6

Contents

1. Designs, constructions and propeller arrangements 4

1.1 Tricopters 4

1.2 Y6- copter 8

1.3 Quadcopter, hexacopter, octocopter 9

1.4 Propeller arrangements for quad-, hexa- and octocopters 11

2. Semi scale models 18

2.1 Semi scale models of person- carrying systems. 19

2.2 Hiller's aerial sedan 22

2.3 Flying jeeps and the X-19 from Curtiss-Wright 26

3. More ideas 30

3.1 An air cabriolet 30

3.2 United starship enterprise 31

4. Frame construction 35

4.1 Crash- safe or with predetermined breaking point 35

4.2 Construction of a DIY-store frame 38

4.3 Safety 42

5. Literature 44

1. Designs, constructions and propeller arrangements

Most flying multirotor systems are equipped with four propellers. But there are other designs with fewer or more rotors. In order to remain general, however, they are no longer called quadcopters, but rather multicopters. The former have the number 'four' in their name. The multicopters have different advantages and disadvantages compared to the quadcopters. Discussing these and also their propeller arrangement is the purpose of this chapter

1.1 Tricopters

If one were to ask how many speed-controlled propellers are actually needed to keep a multi-rotor system stable in the air, the answer would be three.

Figure 1: Tricopter with aluminum frame, photo: Karl-Heinz Wolf.

When using an odd number of propellers however, the same number of left- and right-rotating propellers can't be used. Thus, the yaw torque must be compensated in a different way. The tricopter constructors usually use from the beginning only right-turning propellers. Since these belong to the standard range in model construction, they are present in much greater choice in the market. One is then not forced to choose a propeller that has a right- and left-turning design.

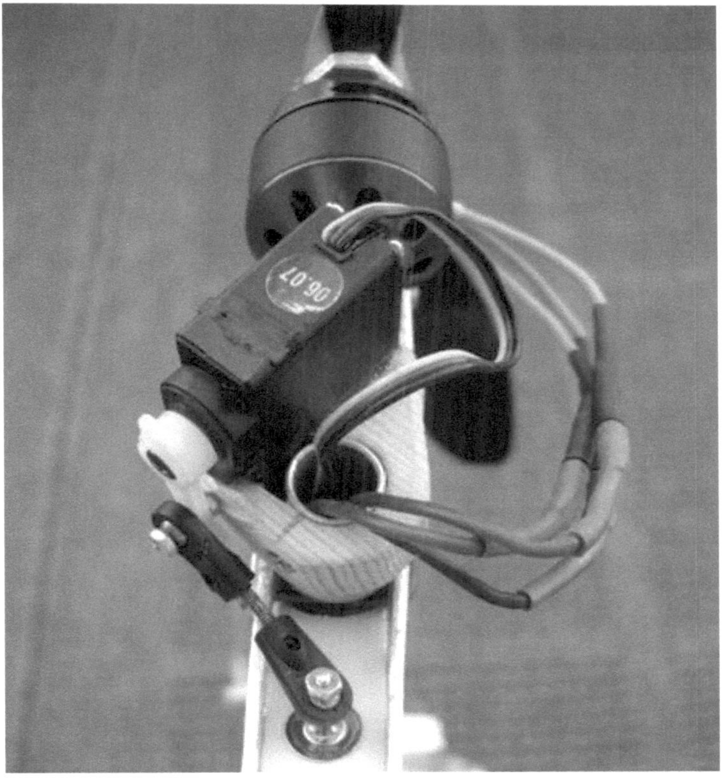

Figure 2: Pivoting boom, photo: Karl-Heinz Wolf.

The consequence is that a drive must be pivotally designed. In this way, a compensation of yaw torque or, if desired, a rotation around the yaw axis can be achieved. The tricopter in figures 1 and 2 was built with an aluminum frame.
The control of the angle φ in the nick and roll axis is a bit more complicated. The two axes are no longer decoupled. Imagine that the pivoting boom of the model is 'rear'. Then the other two booms are 'front' (actually, front-right and front-left). In order to steer the nick axis, both front drives need to act together, whereas in the rear there is only the 'rear' drive. To control the roll axis, the drives 'front-left' and 'front-right' operate.
The two front booms therefore have a double function: They must simultaneously stabilize the nick and the roll axis. To control the nick axis they both produce at the same time more or less thrust, whilst to control the roll axis one drive produces more thrust and the other less. Of course, with a programmable electronic control system this can be easily implemented.

Figure 3: Tricopter with carbon booms, photo: William Thielicke.

Figure 4: pivoting boom, photo: William Thielicke.

Figures 3 and 4 show a further tricopter with a pivoting boom. The frame and the boom were realized very beautifully and shaped with carbon.

Advantages of tricopters:

- *They need only three motors, propellers, booms and brushless controllers, which means a corresponding reduction in price.*
- *You can use right-turning propellers, of which there is a much greater choice in the market.*

Disadvantages of tricopters:

- *One axis must provide pivoting.*
- *The control is more complicated because the nick and roll axes are no longer decoupled.*

1.2 Y6- copter

The system shown in figure 5 is actually also a tricopter. At least it has three booms. The difference to the tricopters presented above is that not only three but six motors and propellers are mounted. ‘Y’ stands for ‘three booms’ and ‘6’ stands for ‘six motors and propellers’. As the picture shows, a left- and right- rotating propeller is always mounted to the bottom and the top.
The control in the nick and roll axis is the same as explained above. The two rotors of an axis both rotate at the same speed. In contrast to above a pivoting axis is now no longer required. For yaw, one of the two propellers of an axis simply turns faster and the other slower. The downward propellers, however, are a little more vulnerable to breaking on landing. In addition, the price advantage of the above-described tricopter here no longer applies. Even more motors and propellers are needed than with a quadcopter. Nevertheless, the Y6 configuration is a special form.

Figure 5: Y6- copter.

1.3 Quadcopter, hexacopter, octocopter

Many technical systems are designed as redundant for safety reasons. This means that individual components can fail without affecting the overall function. Hexa- and octocopters are an excellent example of this.
If during the flight with a quadcopter a brushless motor or controller fails, it unavoidably sags the boom, and a rollover is the fatal consequence. Auto rotation, which is often seen in emergency cases with helicopters, is not possible here. The small moment of inertia of the propellers vitally important for controlling is in this case just a disadvantage.
These failures are very rare and when they happen it is almost always due to a wrong combination of motor and propeller. A camera pilot is often hovering around with a heavy and expensive camera. It is understandable that he fears this terrifying scenario. For this reason, hexacopters with six or octocopters with eight propellers are built.

Figures 6 and 7 show a hexacopter and an octocopter. They consist of two components:

a) The frame with the motors and controllers (mounted underneath the motors)

b) The exchangeable electronics dome, with all components

With the hexacopter, the bottom-mounted camera can be seen very clearly. This enables images from all perspectives, which can be generated without interfering propellers. The landing skid offers protection against damage.

How many drives can fail with the model still landing safely also depends on whether they all fail on the same side or whether the failures occur symmetrically. In any event, a reasonably safe landing should still be possible if one drive fails.

Apart from the redundancy, the hexacopter and octocopter – assuming the same motors and propellers – generates in total 6 / 4 = 1.5 and 8 / 4 = 2 times more thrust than the quadcopter.

Figure 6: Hexacopter with camera, photo: Johannes Thor.

Advantages of hexacopters and octocopters:

- *Due to the redundant design of the motors and propellers, at least one drive can usually fail in flight without the copter suffering damage.*
- *With the same motors and propellers more thrust will be generated in total than with a quadcopter.*

Disadvantages of the hexacopters and octocopters:

- *The greater number of motors, propellers, booms and brushless controls corresponds to a higher price.*
- *As more components are needed, the construction, putting into service and maintenance is more complex.*

Figure 7: Octocopter, photo: Melanie Plischka.

1.4 Propeller arrangements for quad-, hexa- and octocopters

For quadcopters, the so-called '+' or 'x' configuration is used. This is shown in figure 9. Due to the better feasibility in the combination with cameras, the 'x' arrangement is usually implemented. Here, 'in front' is marked with the suggestion of an arrow. As can be seen from this, the booms of the quadcopter represent a '+' or an 'x'. With the even larger number of rotating propellers of the hexa- or octocopters, one can imagine many different ways in which they can be arranged.

Figure 8: In reality a roll bar would probably make sense.

Figure 9: ‚+'- and ‚x'- arrangement of quadcopters.

Figure 10 shows a possible arrangement of the propellers in a hexacopter. These are arranged in a ring here and each arm is connected to the center, in which the control electronics is usually located. In principle, it would also be possible for the motors to sit on an outer carbon fiber ring and there are only three or four booms which connect it to the center. A distinction is now made again as to whether the arrow, which indicates the front, points into the direction of an arm or whether it points into the direction between two arms. Here it is already a little difficult to call this either a '+' or an 'x' arrangement, since it does not describe any of these letters. The left part of the picture could also be referred to as an 'I' arrangement and the right one as a 'V' arrangement. You can leave these terms alone and just let the pictures speak for themselves. So that the torque around the vertical axis can be compensated again, a total of three left-hand and right-hand rotating propellers must be installed. These are then placed alternately in the circle.

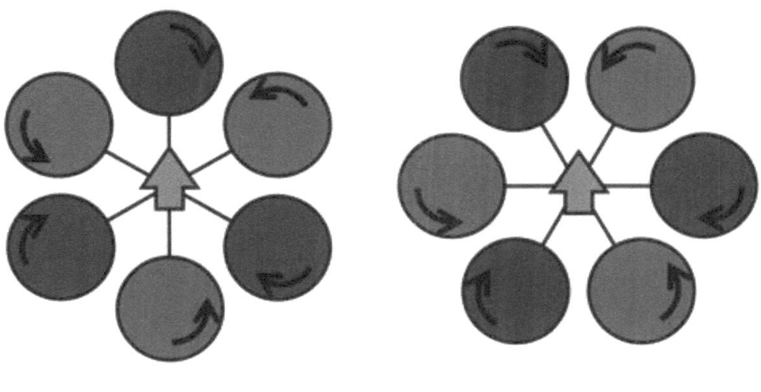

Figure 10: Circular propeller arrangement at hexacopters.

Figure 11 shows the same with octocopter propellers arranged in a ring. Here, too, the arrow points in one arrangement in the direction of a boom and in the other between two booms. Since 2 x 4 is known to result in 8, octocopters are actually twice quadcopters. But arrangements according to figure 12 are built somewhat less frequently in practice.

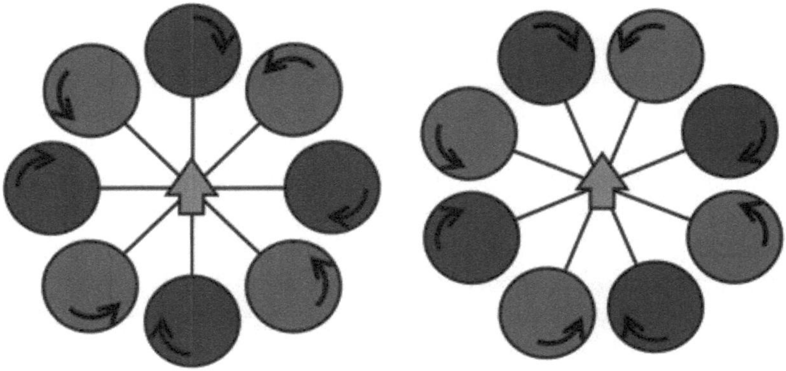

Figure 11: Circular propeller arrangement at octocopters.

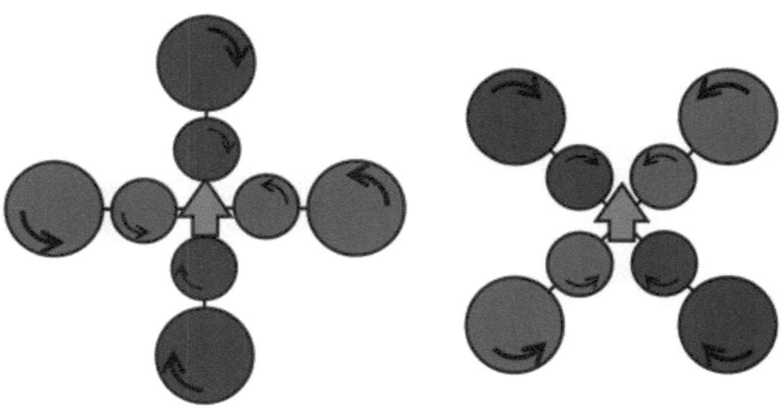

Figure 12: Further possibilities for the propeller arrangement in octocopters.

Here, two motors and propellers sit on one boom. So you can save some weight in the frame structure and the '+' and 'x' arrangement can also be clearly seen here. For geometric reasons, the inner motors and propellers can also be made somewhat smaller than the outer ones. The motors can then be controlled in the same way as with the quadcopters. The two drives on the same axis then each receive the same control signal. A closer look at this system shows, however, that the entire size of the system cannot be optimally converted into propeller surface. The inner propellers take up some space and there are larger empty spaces between the outer ones.

Hexa- and octocopters are sometimes built in the so-called 'H' arrangement as shown in figure 13. All drives are arranged lengthways on two sides. With a little imagination you can recognize the letter 'H'. Here, too, the propellers are typically mounted alternately to the left and right. Figure 14 shows such an octocopter in flight. The propeller arrangement was selected in figure 13 on the right. Every two propellers lying next to one another have the same direction of rotation and their motors are also controlled with the same signals.

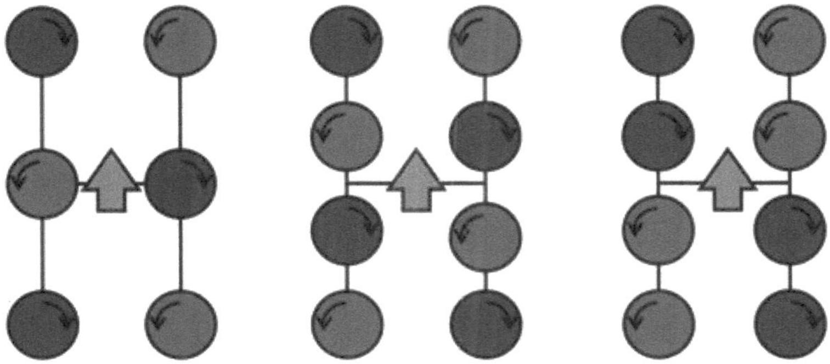

Figure 13: ‚H'- arrangement at hexa- und octocopters.

Figure 14: An octocopter in ‚H'- arrangement during flight.

Figure 15: Carbon frame of an octocopter in ‚H'- arrangement.

In practice, the most frequently built multicopters are quadcopters, followed by octocopters for larger loads. The hexacopters are a little less common and the tricopters are even less common. You have to consider that the cost of building a multicopter increases with the number of propellers. That is why systems with twelve, sixteen or even more propellers are also conceivable, but they are usually only built as individual experimental models.

2. Semi scale models

All in all, quadcopters would be much better accepted among model-makers if they were a little more colorful. Many pilots say, however, that all the work building a depron body would be destroyed in a rollover. However, anyone who flies quadcopters today notices that the technology has advanced so far and so safely that (semi-)scale models can be built without problems in this regard. The only problem that arises is that there are not as many flying originals.

In addition, better visibility and orientation is also ensured with such a model. Many pilots can only fly their quadcopter a few meters away from themselves, because otherwise they can no longer see where the front of their model is. A depron construction gives a completely new feeling of flight, and flight radiuses are also possible which certainly come close to those of normal model airplanes.

With model aircraft, the fuselage is responsible for the correct fit of the wings; that means 'form follows function'.

This does not apply here. The bodies – made of plastic, depron, painted paper or carbon – have no special function. They are actually only ballast. But exactly this may be interesting. Due to this fact, the imagination knows no limits.

This chapter describes semi-scale models. In order to build semi-scale models, a very decisive prerequisite is required: there must then also be person-carrying systems as a counterpart to these models. And they actually exist, the very few person-carrying quadcopters that were actually built or at least planned.

2.1 Semi scale models of person- carrying systems

The Oehmichen No.2 was the world's first quadcopter. Etienne Oehmichen started in 1922 with what was probably the first vertical take off and landing system. As with today's systems, four horizontal propellers provided buoyancy. The helicopter swashplate had not yet been invented, so alternative ways of setting variable rotor pitches had to be found. In addition, the rotors on the Oehmichen No.2. were elastic and could be deformed with rope cables. There were also smaller vertical propellers, which on the one hand provided propulsion and on the other hand served to compensate for the torque around the vertical axis. This invention was later taken up again in the development of the tail rotor of helicopters.

In the years that followed, things went quiet about quadcopters, since the development of helicopters through to series production was the more promising approach. In retrospect, it still is today.

The helicopter pioneers Stanley Hiller and Igor Sikorsky knew that helicopters, from the energetic point of view, would never match airplanes. They would need more power to keep themselves in the air. Nevertheless, both developed these aircraft up to today’s perfection.

They also knew that with helicopters you can solve tasks that are impossible to accomplish with airplanes. A helicopter can stop at any point in the air. Therefore, it has become an indispensable tool in many fields.

In the 1950s also the vision of individual traffic with a four- rotor helicopter, an early quadcopter, emerged.

Figure 16: A model of private transport with quadcopters.

The vision

In 1957, the July issue of the magazine 'Popular Mechanics' contained an article about 'Hiller's Aerial Sedan, Your Flying Car for 1967'. It was announced that almost everybody could buy such an aircraft in 10 years. The functioning was clearly explained, namely that for forward flight the speed-controlled front propellers would rotate for a short time slightly slower than the rear. Braking would be achieved by a slower rotation of the rear propeller. There were at that time some four-rotor prototypes for the army, such as a flying jeep or the X-19 by Curtiss-Wright.

If one wanted to go into this market, then it was clear to everyone that these VTOLs (Vertical Take Offs and Landings) would have to be improved also constructively.

Quadcopters with their vertically arranged rotors represent a considerable safety risk. One must only remember that passengers would have to get in and get out between turning rotors. Furthermore, in the case of private traffic, it must be possible to start and to land safely everywhere, even near to people. The solution to this is to integrate all rotors completely into the bodywork.

Figure 17: Vision 1957, private transport with quadcopters.

The disillusion

Of course, they brought the flying jeeps into the air as well as the X-19. But they soon had to recognize that the technology was just not mature enough to build something like an air car for everyone. It was also missing the microprocessor technology, which is standard in today's quadcopter models. Together with accurate and drift-free sensors and powerful motors, they allow robust stabilization for the first time.

Therefore, in those days only experienced pilots could steer these prototypes. Various setbacks and crashes finally led to the fact that this vision remains unfulfilled to this day. The Aerial Sedan, the flying car for 1967, was never built. To date, individual transport takes place largely on wheels and on the ground. The third dimension is, as already at that time, reserved only for a few.

2.2 Hiller's aerial sedan

Nevertheless, an aerial sedan can be built as a model. Figure 19 shows this in 'semi-scale'.

As a basic platform, a commercially available flight controller is used. The motors and propellers are Hacker A20-22-L and 10" x 4.7". This combination generates with all four drives together a little more than 2kg (20N) of thrust. This is enough to build a nice model with a total weight of about 1kg, and to maneuver it safely in the air.

Figure 18: Carbon construction and safety rings for the 'aerial sedan'.

The basic structure is realized with carbon fiber tubes. The control board, the receiver and the four brushless controllers are all placed on the center plate. This is made out of the sandwich with two 1mm carbon fiber panels and an aluminum cross of two 10mm square tubes. This can, on the one hand, take the 8mm carbon tubes of the booms and ensure, on the other hand, that the electronics will not be damaged in a crash.

Below are four 6mm nylon screws, which provide for a sufficient distance from the ground. The quadcopter stands on the wing bolts on the bottom. Here is also enough space available to include a 3S-2100mAh-LiPo accumulator. The electronics are also protected from the top, using nylon bolts and pieces of carbon fiber. Since the length of the 'Aerial Sedan' is larger than its width, two longitudinal tubes are attached. At their ends they take the drives. Two crossbars complete the construction. They are responsible for ensuring that, in operation, the shafts of the motors do not twist against each other. Figure 19 shows the construction.

Unfortunately, the finish of painted depron often looks a little like bricolage. Therefore, a new method with the use of a self-adhesive foil is followed. Its adhesion to the depron is not quite as good as if it were glued to wood. If, however, the foil is sitting right the first time and is not removed for correction purposes, then it glues quite well and gives the models the same look as if it were disguising any other material. With this combination, a very light and elegant body can be built. Unlike with most other flying models, it has no supporting function. It is just the carbon tube construction which provides for that.

The hood is built more conventionally from 5mm x 5mm balsa strips. The roof consists of covering paper, which is finished with red spray paint.

Figure 19: Aerial sedan as a semi scale model.

Figure 20: A look into the propeller housing.

Safety

What is true for a (never built) original applies also for the model. When the company Silverlit brought the X-UFO to the market in 2005, the relatively soft 8" propellers were still protected with EPP rings. Today quadcopters are often operated with larger motors and 10" propellers. These are also sometimes relatively heavy. Flying quadcopters, however, must not be a safety risk for spectators and pilots. That is reason enough to treat this issue deeper. In the present model, rings of 50mm x 930mm x 0.5mm carbon fiber are mounted as a propeller guard. The extra weight of all the rings is 120g, which makes up about 10% of the total weight of the model. It would be great if this 10% was used in each of today's flying quadcopters for safety!

Figure 21: The model of an aerial sedan in pink.

2.3 Flying jeeps and the X-19 from Curtiss-Wright

At that time, however, a few person-carrying quadcopters were actually built in real. There were four-rotor prototypes for the army, for example a flying jeep with the designation VZ-7AP or the X-19. Both aircraft were developed by Curtiss-Wright. The X-19 was further developed from the prototypes named X-100 and X-200 in the early 1960s. It had two wings of similar size attached one behind the other. Four engines, each with a propeller, were attached to the end. The specialty was that they only worked as quadcopters during take-off. In flight they should be swiveled 90° forward. The X-19 should be able to take off and land vertically, but otherwise behave like a normal fixed-wing aircraft. This would have combined the advantages of a VTOL and those of a fixed-wing aircraft. In particular, the forward flight as a completely normal aircraft could have been made much more efficient from an energetic point of view.

However, the control of the four engines and, in particular, their 90 ° pivoting during flight resulted in great technical difficulties for the engineers. In total, only two prototypes of the aircraft were built. After considerable difficulties, the flight tests were canceled in 1965. The semi-scale model shown in figure 22 with an all-up weight of around 1200g has no swiveling rotors. But it is equipped with a GPS-enabled quadcopter.

Figure 22: Semi- scale- model of a Curtiss- Wright X-19.

2.4 Private transport with quadcopters from today's perspective

One may wonder whether it would be possible from today's perspective to extend the private transport in the third dimension using quadcopters. Basically the idea is still great. It would change the world at least twice as much as the invention of the automobile itself.

There is technically no reason today why the ‘Aerial Sedan’ should not be built for real, for use as private transportation. The vision would be technically feasible. The controller could be designed with assistance sensors so that anyone with a driver’s license would be able to maneuver it safely through the air.

However, there are other aspects that must be considered. The dimensions of a modern automobile would be the graduated measuring rod by the construction. With a width of about 2m, the rotors could therefore have a diameter of 1m each. With a lightweight construction, a take-off weight of 500kg may be achieved. Applying a little theory, one finds that more than 100kW would be needed just to hover. Since a quadcopter still requires reserve power for the regulation and forward flight, a 250kW motor would be sufficient.

Physically, the specific thrust increases with a larger rotor area. Therefore, the rotors of helicopters are indeed very large. So if one could assume a greater rotor diameter than 1m, the power/thrust ratio would be much better. However, as in the 21st century energy is increasingly developing into a resource problem, it appears to make little sense to use the quadcopter for private transport.

Another question would be to clarify what would happen if the drives fail. A helicopter can land safely under these circumstances with autorotation. A quadcopter, which owns speed-controlled rotors, has propellers with small inertia. The regulators must be able to change the speed quickly in order to provide a stable flight. Autorotation is therefore not possible. So if a rotor fails, a rollover is the fatal consequence. This is why in today's real projects more than four propellers are implemented because of the redundancy

A car on four wheels doesn't have the above problems. If the drive fails, it simply doesn't move anymore. The energy balance is much better, since no power is needed for hovering. All energy can be spent on forward movement.

In summary, it must be said from the current perspective: The quadcopter is for the time being reserved for model flying and unmanned flying. There, its potential is very large. From an energetic point of view and because of the safety considerations in the event of the failure of the rotors, it does not make sense for use it for transportation of persons.

3. More ideas

For better orientation in the air, it is not absolutely necessary to implement semi-scale models of quadcopters that were once built or planned. Of course, the development of the quadcopters also invites you to come up with fantasy structures. This sub-chapter is intended to take account of these many different forms.

3.1 An air cabriolet

Figure 23: Larger models look almost like real.

The model of an 'air cabriolet' (in reality, for now, non-existing) shown in figure 23 consists of 3mm depron. To allow a good assembly, the middle of the interior was strengthened with some balsa strips.

In speed flight there is still a bit of 'function' felt. At higher speeds, the rigid rudder helps a little to keep the forward direction. The weight is 75g in total, including a model pilot. In the case of an 800g copter this is barely noticeable. Basically a quadcopter with a thrust–

weight ratio of 3:1 can take a little more than half its own weight. Nevertheless, every gram in the depron body should be paid attention to.

3.2 United starship enterprise

Figure 24 shows the 'United Starship Enterprise' from the 1960s television series 'Star Trek'. The entire structure is about 300g, 120cm long and 50cm wide. The 'warp gondolas' which can be seen in the back are made of painted and rolled paper, while the front part is made of 3mm painted depron with a diameter of 50cm. The whole structure is held together with carbon rods. It is a little susceptible to wind gusts, but was already successfully used at model helicopter flight events with moderate wind.

The models in figures 25 and 26 are smaller, but also made from 3mm depron. The 'Galileo' and the 'Work bee' are also models of small space ships from 'Star Trek'. With just 25g weight, they also fit quadcopters with 30cm frames and 30g motors.

The rocket in figure 27 is made of painted paper. The total weight is only 25g. The pilot body is also Depron. When the quadcopter is attached for forward flight, the rocket flies almost in horizontal position. That looks pretty good. Due to the aerodynamic shape, it is not susceptible to wind. Also this construction can easily be used with 30cm frames and 30g motors.

The UFO shown in figure 28 has a frame made of 2mm-thick carbon rings. They are held together with balsa strips. Painted paper gives the model the unusual appearance. The weight is 75g with a diameter of 1m. A peculiarity with this construction is that the quadcopter with a 30s frame and 30g motors disappears completely. This is only possible because it is open above and below and therefore the propeller airflow is not disrupted.

Figure 24: Semi-scale? The U.S.S. (United Starship) Enterprise will in any event be launched in the year 2245. 'Who owns one, doesn't need to believe'

Figure 25: Galileo, the shuttle of U.S.S. Enterprise.

Figure 26: The 'Work bee' is used for the construction and repair of star ships.

Figure 27: The rocket is made of painted paper; the weight of the structure is 25g.

Figure 28: UFO in European airspace?

4. Frame construction

The frame carries all components, motors, propellers, control board, brushless controllers, battery and receiver. It is often underestimated. It has the same function as the fuselage in model airplanes.

With a quadcopter which does not show the required behavior in flight, the frame is very often the cause of the problem. This may be because the drives are not sufficiently rigidly mounted, the control electronics with the sensors vibrates, the antenna cable is wrapped around the rotary motors, etc. It will also be pointed out here what one has to consider with the purchase or construction of the frame.

4.1 Crash- safe or with predetermined breaking point

Depending on the field, the author would distinguish between the two frames in the title. For a speed flight oriented pilot, a crash-safe framework would be recommended. Such frames can be bought in several online shops. They are usually drawn with a CAD program; sometimes even the forces acting to the material are simulated and tested. Often, a material is used which until recently was reserved for the professionals: carbon. It is characterized by being very strong but also very light. So it is ideal for these purposes.

Figure 29 shows such a framework. The electronics and battery are protected by a metal case, whereas the boom consists of a sandwich with balsa wood glued between two carbon plates. The brushless controllers are mounted under it. The price of crash-safe frames is higher.

Frames can also be less crash-safe, if they are at least rigid. In this category are the so-called do-it-yourself shop frames. They get their name because you can find almost all the components for the construction in the DIY home-improvement store. The hardware of such frames, however, should be strictly engineered so that if something breaks, no expensive electronics or drives must be replaced. This means that they need a predefined breaking point.

Figure 29: Crash-safe frame.

Frame size

The frame sets the size of the quadcopter. An important indicator is the distance between two opposing motor shafts. There are only geometric constraints that determine them. Care must be taken to ensure that the propellers can rotate freely. But it is not necessary that for a smoother balance, for example for a camera flight, that the distances of the motors are larger. It is the quality of the sensor measurement, the quality of the motors and propellers and their combination, and the controller parameters, which determine this. Simple rules can be created for the frame size as well:

8" propellers will need 30cm to 40cm distance between two opposite motor shafts

10" propellers will need 40cm to 50cm distance between two opposite motor shafts

12" propellers will need 40cm to 50cm distance between two opposite motor shafts

14" propellers will need 60cm to 70cm distance between two opposite motor shafts

It always also depends on where the brushless controllers are arranged. With most crash-proof purchase frames, the booms are so big that there is also room for the controllers. So in this case the center will be made smaller, since it only needs to carry the control board, the receiver and the battery. The distance between two opposing motor shafts is then smaller. With a DIY-store frame, the booms are usually made of aluminum or carbon tubes, which may not carry any additional electronics as they would otherwise not be sufficiently protected. Then the center has to be designed larger, as also the brushless controllers have to find a place (see figure 31).

If more than four propellers are arranged, for example six or eight, as is the case with hexacopters or octocopters, the geometric arrangement must be checked in order to be able to make statements about whether the motor distance is sufficient in each case. In order to avoid problems caused by turbulence at the end of the propeller, it is advisable to always keep a minimum distance of 4cm to 5cm between two adjacent propellers. In some cases even larger multicopters than those with 14" propellers are built. There are direct drives for quadcopters on the market today, which can drive propellers of up to 18" or even more.

4.2 Construction of a DIY-store frame

The hobby constructor of a quadcopter needs to buy electronics, motors and propellers anyway. Thus, only the frame has got a real potential to save money.

For this reason, the construction of a relatively cheap DIY-store frame is shown, which is very rigid for all sizes and has a breaking point in the case of a crash. So it protects the electronic components and, moreover, it looks pretty good. The dimensions are about for motors with 60g and 10” or 12” propellers.

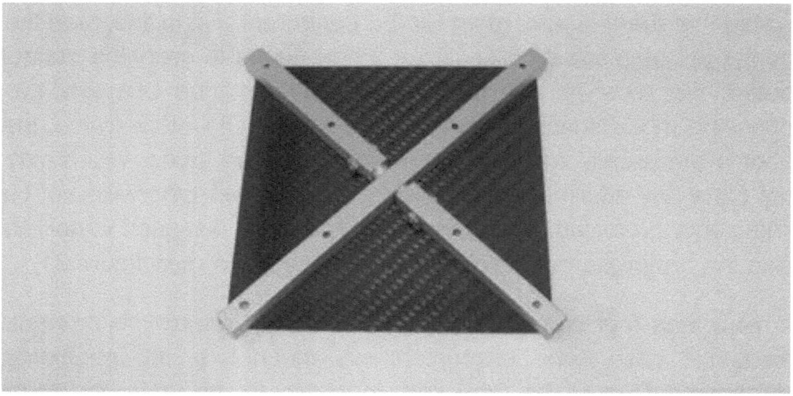

Figure 30: center plate, aluminum cross.

The center plate of 130x130mm must carry the expensive electronic components and the accumulator. It is therefore important to ensure that it does not suffer any damage in a crash. So, first, an aluminum cross is built according to Figure 30. The square tube has the outer dimensions 10x10x185mm and the interior dimensions 8x8 and can carry a carbon tube of 8mm diameter.

In the middle, an 8mm hole was drilled to hold an 8mm aluminum tube. This is attached to two 10x10x85mm square tubes and screwed together. 10mm from each edge, a 3mm hole is drilled. Later on, the booms will be screwed into them.

Now two 130x130x1mm carbon plates are cut, which are bolted to the aluminum cross with M3 nylon screws. The distance of the holes from the center is 50mm each. Since the carbon plates are relatively expensive, 0.8mm aluminum plates can be used instead. However, there are then some restrictions, as a higher weight and a lower strength is the result.

The accumulator is on the bottom side and is held with M6x40mm nylon screws at a distance from the base. For the protection of the electronics, carbon pieces are used together with the same nylon screws. As spacers, carbon tubes 8mm in diameter are fitted. Figure 31 shows this.

The booms are made of 8mm carbon tubes; alternatively and with the same restrictions as above, also aluminum can be used. Many DIY-store frames whose booms are designed with tubes are not rigid enough and the drives can quite easily be twisted. The manufacturers of electronics warn, as in this case, that no reliable control of the system (no reliable balancing) is guaranteed.

The version presented here solves this problem. At the end of the carbon tubes are again 10x10x70mm square aluminum tubes for the drive mount, bolted with M3 screws. In addition, two booms are plugged together using other carbon tubes (4 to 6mm) and smaller aluminum square tubes. This is for those who are flying in 'x' configuration (see also figure 9 again). If flying in the '+' configuration, for reasons of symmetry all four booms should be connected with four 4mm carbon tubes. Of course this creates an even better rigidity with a simple and relatively lightweight solution.

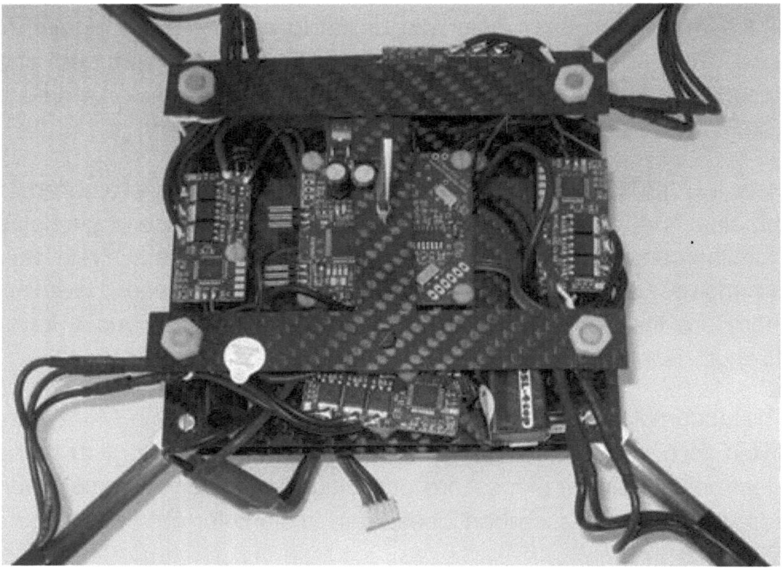

Figure 31: Center plate with electronics and brushless controllers, top view.

This additional connection provides on one hand for sufficient rigidity and ensures that the drives can't move against each other, and on the other hand also ensures that all motor shafts are vertical. Figure 32 shows a completely screwed boom.

In a crash, this framework is certainly not as crash-proof as a purchased frame, which is designed for speed flight. However, you will always have the 8mm carbon booms to break and furthermore exactly where they exit from the square tube of the middle piece, so there is a breaking point. This is very desirable for two reasons: The expensive electronics lie safely on the unbreakable center plate and the deformation energy of the crash does not act on the motors and propellers. Most importantly, only a relatively cheap part, the 8mm carbon tube, which is attached with two screws, needs to be replaced. In five minutes the damage is repaired on the airfield and the flight can continue. From 1m carbon tube more than one replacement tube can be manufactured. Depending on the size of the quadcopter, their length is from 15 to 25cm.

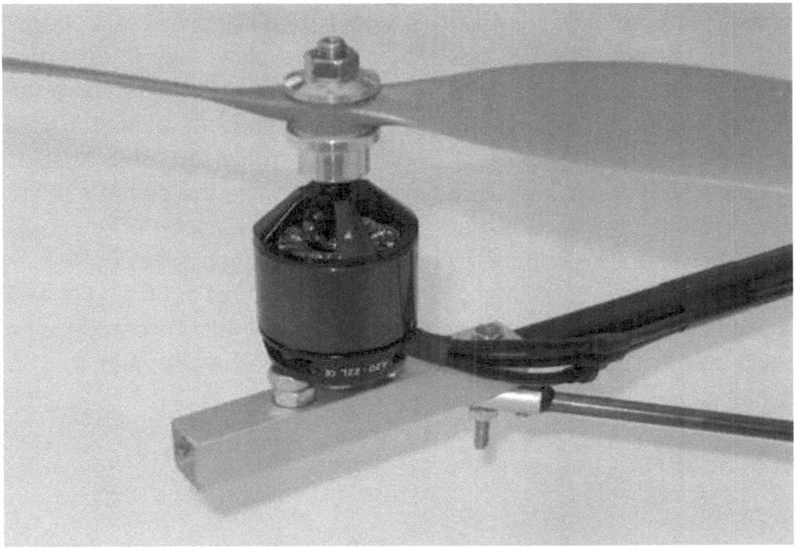

Figure 32: Completely screwed boom with a 60g motor.

Adaption to smaller systems

Smaller propeller diameters are less likely to be realized with self-made multicopters, these sizes are reserved for RTF quadcopters. They are available in all sizes between about 1" and 7", and since all materials, including motors and propellers, are a few sizes smaller, they can be manufactured cheaply and also sold in large numbers. These systems then usually have cameras and other electronics built in that are precisely matched to the total weight. It has to be that way, because the permissible payload with these systems is quickly reduced to significantly less than 100g or, depending on the size, even only 10g or even less.

Such small payloads are not interesting enough for self-construction systems. In principle, however, it would be quite possible to adapt the concept presented above with a middle plate, aluminum cross and carbon tubes to smaller sizes. Carbon tubes with 6 mm or 4 mm diameter and corresponding square tubes made of aluminum can be processed well and are also readily available. With smaller systems, you would have to pay less attention to the stability of the carbon tubes, as they are short and therefore stiff enough. You could also do without aluminum mountings and instead produce lighter components with 3D printers. Since the stability of the frame is well guaranteed with such small multicopters even without carbon, one could also imagine making this completely self-made with the 3D printer.

4.3 Safety

The technology is interesting, but it is also important to consider safety. As with all helicopter-like aircraft, quadcopters also have propellers which are mounted in the horizontal. On top of that, the used propellers are often very hard. Accidents caused by uncontrolled quadcopters falling on people could have dire consequences.

Safety is one of the basic components – it is not optional! Opportunities for this are rings made of styrofoam or as pictured in figure 33, carbon fiber. Here, 0.5mm-thick plates were used. In the concrete example they can be realized with a weight gain of 120g.

Figure 33: Safety ring made of carbon.

Figure 34: Quadcopter with safety rings.

5. Literature

[1] Schulte, Hinrik. *Das Depron-Buch: Modelle, Bautipps, Fliegen*. Verlag für Technik und Handwerk, 2007.

[2] Traxler, Wolfgang. *Indoorfliegen für Einsteiger:• Modelle• Antriebe• RC-Komponenten• Bautipps• Flugtechniken*. Verlag für Technik und Handwerk, 2010.

[3] Buchi, Roland, et al. "A remote controlled mobile mini robot." *MHS'96 Proceedings of the Seventh International Symposium on Micro Machine and Human Science*. IEEE, 1996.

[4] Moderne Fernsteuerungen für RC-Flugmodelle, Manfred-Dieter Kotting, Verlag für Technik und Handwerk GmbH, ISBN: 978-3-88180-780-7

[5] Büchi, Roland. "Brushless-Motoren und–Regler. 1." *Auflage, Baden-Baden: Verlag für Technik und Handwerk neue Medien GmbH* (2013).

[6] Leishman, J.G. (2000). "Principles of Helicopter Aerodynamics". New York, NY: Cambridge University Press. ISBN: 978-0-52185-860-1

[7] Hoffmann, G.M.; Rajnarayan, D.G.; Waslander, S.L.; Dostal, D.; Jang, J.S.; Tomlin, C.J. (November 2004). "The Stanford Testbed of Autonomous Rotorcraft for Multi Agent Control (STARMAC)". In the Proceedings of the 23rd Digital Avionics System Conference. Salt Lake City, UT. pp. 12.E.4/1–10.

[8] Büchi, Roland. *Radio control with 2.4 GHz*. BoD–Books on Demand, 2014., ISBN 978-3732293407